UFAW ANIMAL WELFARE RESEARCH REPORT NO.2

THE DISTURBANCE INDEX:

A BEHAVIOURAL METHOD OF ASSESSING THE SEVERITY OF COMMON LABORATORY PROCEDURES ON RODENTS

by

R J BARCLAY

W J HERBERT

T B POOLE

Published by

UNIVERSITIES FEDERATION FOR ANIMAL WELFARE

8 Hamilton Close, South Mimms, Potters Bar,

Hertfordshire, EN6 3QD, England.

ISBN 0 900767 51 0

AUTHORS' AFFILIATIONS

R J Barclay, ICI Pharmaceuticals, Mereside, Alderley Park, Macclesfield, Cheshire, SK10 4TG, formerly of the Animal Facility, Medical School, University of Dundee.

W J Herbert, Laboratory Animal Science Association, 20 Queensbury Place, London SW7 2DZ, formerly of the Medical School, University of Dundee.

T B Poole, Universities Federation for Animal Welfare, 8 Hamilton Close, South Mimms, Potters Bar, Herts EN6 3QD.

The research was carried out at the Animal Facility, Medical School, University of Dundee.

ACKNOWLEDGEMENTS

The authors wish to express their gratitude to:

The **Universities Federation for Animal Welfare** for its financial support;

Dr Alastair Todd of the Department of Pathology, Ninewells Hospital and Medical School, for the loan of a time-lapse camera and free film, without which the early work which established the method as viable could not have been carried out;

Mr Emil Andersen of the Department of Physiology, University of Dundee, who designed and constructed the movement detectors at minimal cost;

Dr Simon Ogston of the Department of Statistics, University of Dundee, whose interest in the project led to the statistical validation of disturbance indices;

Dr D J K Balfour of the Department of Pharmacology, University of Dundee, whose suggestions for the tests with drugs enabled a clear demonstration of the utility of the indices to be made;

Dame Olga Uvarov whose continual interest, suggestions, stimulation and prodding was vital to the completion of the project.

Finally the authors wish to express their special gratitude to **Miss Samantha Tozer** for her skill and patience in assembling and concatenating various parts of numerous disc files received from three different sources, not to mention pieces of semi-illegible hand written script.

FOREWORD

The **Animals (Scientific Procedures) Act 1986,** in specifying the conditions under which a personal licence shall be issued to a scientist, states that "... the holder shall take precautions to prevent or reduce to the minimum consistent with the purposes of the authorised procedures any pain, distress or discomfort to the animals to which those procedures may be applied...", Section 10-(2)(a).

This condition for granting a licence appears to make the assumption that pain, distress and discomfort can be measured in animals. While no objective method has yet been devised to accurately assess the extent to which an animal may be suffering, it can generally be inferred from observed departures from the normal behavioural repertoire. The extent of behavioural change, whether it is an increase or decrease in a particular activity, can be assumed to relate directly to the severity of pain, distress or discomfort which the animal is experiencing.

This Report describes a method of scientifically assessing the degree of disturbance to a rodent's exploratory behaviour caused by subjecting it to an experimental procedure.

The method used allows different procedures to be compared and generates a statistic which indicates the extent to which the animal's behaviour has been changed. The statistical significance of this value, which has been termed the **Disturbance Index** or **DI** for short, can be tested. No complex apparatus or observational techniques are necessary to generate this index, which can be calculated in any laboratory using simple activity meters and event recorders. Scientists can now begin with some confidence to start assessing the disturbing effects of different procedures on laboratory rats or mice. There is little doubt that the method will be applicable to other species of rodent and to a wide range of experimental techniques.

UFAW hopes that, in future, scientists will utilize the methods described in this Report to investigate procedures the relative severity of which are unknown. The knowledge thus acquired should allow them to select for future use those procedures which achieve the desired experimental results but with the minimum disturbance deviation from the norm. The probability then, is that they will have reduced to the lowest possible level the pain, distress or discomfort to the animals.

Roger Ewbank
Director, Universities Federation for Animal Welfare

1. INTRODUCTION

1.1 Assessing pain and distress in animals

Pain and distress in animals are customarily assessed by reference to human experience. The assumption is that any procedure which would cause suffering in a human would have a similar effect in an animal and it is also assumed that the responses of the animal to pain would be similar.

Based on these assumptions, several methods for the quantification of pain and distress in animals have been proposed, see References 1,2 and 3. All of them rely on a subjective grading of symptoms shown by the animals. Such methods are valuable as a guide to the severity of many painful procedures. They are not so helpful where there is a need to compare several means of carrying out a procedure, for comparing two entirely different procedures, or for minor procedures such as simple injections. This is because the grading requires the use of discontinuous scales which make it difficult to measure small differences. Nor can the statistical significance of differences be determined except when variation is large.

More seriously, the underlying assumption of a close correlation with human experience, is flawed. All animals have evolved to fit ecological niches so that sensory appreciation of, and responses to, a given environment vary widely amongst species. Thus ultrasound is readily detected by rodents, though unperceived by humans. The symptoms shown in response to injury may also be different. It would be expected that reactions to pain would include crying out, licking the affected part, changes in facial expression and body posture etc (3). In prey species these responses may be suppressed as they would draw a predator's attention to the injured individual. In such species, responses to injury are likely to be a reduced level of activity or even immobility. Such responses are difficult to detect and score subjectively. The problem of subjective grading is further complicated because, whilst pain or distress usually result in the exhibition of certain types of behaviour, such behaviour may not in itself be symptomatic of distress. What is required is an objective means of recording discomfort and distress.

1.2 The concept of disturbance

This Report describes an approach to the detection of discomfort and distress based upon a detailed knowledge of ordinary animal behaviour rather than upon trying to describe abnormal activity. Laboratory rodents were subjected to routine procedures such as handling, physical examination and simple inoculations. Visual observation of the animals showed that, though no overt signs of distress were evident, even very minor procedures produced deviations from normal behaviour. It is difficult to know whether such deviations can be construed as indicating actual distress; though from the nature of the procedures used some, at least, were probably responses to pain.

The degree of distress experienced by the animal would be expected to be reflected by the extent of changes in the incidence of normal behaviour patterns, whether they be increases or decreases. Accordingly, if changes in normal behaviours could be measured, the degree of upset to the animal could be quantified. The problem lay in finding a normal activity which was measurable and sufficiently replicable to allow for statistically valid comparisons to be made of the effect of experimental procedures upon it.

Certain types of behaviour have a high priority in the behavioural hierarchy such that all members of a species carry them out in a predictable fashion. Amongst the behaviour patterns considered for our purpose were feeding and drinking rates, locomotor activity, sleep patterns and so forth. None of these were sufficiently predictable in onset for the purpose. Exploration of a strange environment was however, a behaviour that could readily be initiated in a rodent and therefore seemed a promising candidate.

If a mouse is placed in an unfamiliar cage it will, for some time, show a high level of investigative and scent marking behaviour. It repeatedly explores all parts of the box, urinates and locates the sources of food, water and possible routes of escape. Most other species explore and mark out new territory in a similar way (4) and humans exhibit analogous behaviour on, for example, entering a hotel room for the first time.

Pilot experiments using a movement detector showed that the number of movements made by a normal laboratory rat, or mouse introduced into an unfamiliar cage was highly repeatable. Moreover this number was independent of the time of day and of husbandry factors. Animals subjected to an injection of saline before introduction to the cage, were found to differ in the number of movements that they made. As such differences were changes in normal behaviour rather than a record of expected responses to distress, it appeared that a wide variety of response to stressors or injury could be accommodated and compared.

1.3 The Disturbance Index (DI)

For some procedures a straightforward comparison of the number of movements made under control conditions and those made by the treated animals on introduction to a new cage was informative. With other procedures the pattern was found to be more complex. In some instances techniques stimulated an increase in activity, while in others there was a decrease or even reduced activity followed by a period of compensation such that the aggregate movements in control and experimental situations were similar. It was also difficult to determine the period over which records should be made, and comparisons between techniques and between species were impossible. Raw data were therefore only useful for plotting graphs which were then subjectively interpreted. It was evident that a numerical parameter, derived from the data, which could be used for statistical comparisons was required.

With the assistance of Dr Ogston of the University of Dundee a Disturbance Index was devised. In its simplest form this is the base 10 logarithm of the result of dividing the experimental values by those of the controls. This index is robust and can be tested statistically so that significant departures from the controls are identifiable, see Section 2.6.

In practice it was found necessary to calculate on a temporal basis several indices for each procedure tested. This is because exploratory activity occurs in distinct phases, see Figure 1, and experimental procedures may affect any or all of these. Moreover, a rebound effect is observed in some cases such that an initial decrement in activity may be compensated for by excess activity later in the observational period. Thus a DI calculated over the whole period might show no significant difference from normal and the substantial initial effect be ignored. Division of the observational period into phases suitable for the species under observation was made on the evidence of an analysis of variance, which was carried out on the total incremental count of movements made by the animals on introduction to a clean cage. For further details, see Section 2.5.

This Report presents evidence validating these indices as a means of measuring the disturbing effects on an animal for experimental and other procedures. To this end, a number of handling, restraint and injection techniques commonly applied to rats and mice were examined. In some instances it was possible to test variations of each method, such as volume of inoculum, and to recommend those that appeared least disturbing to the animals.

The concept of the DI method and preliminary results were first presented by Herbert and Barclay to the Laboratory Animal Science Association meeting held in Bath, England in May 1982.

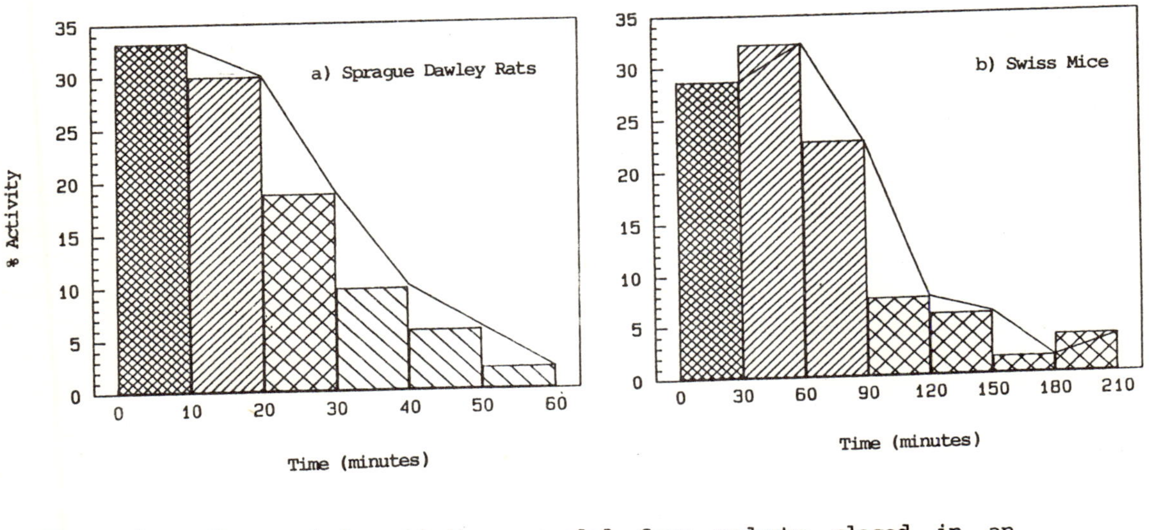

Figure 1 Incremental activity recorded from rodents placed in an unfamiliar cage.

2. MATERIALS AND METHODS

2.1 Animals and diet

Male rats (Sprague-Dawley) and mice (Swiss and C57/Bl6) were used for the experiments. The rats were between 9 and 15 weeks old and the mice were aged between 60 and 90 days at the start of the experiment. All animals used in the experiments were bred at Ninewells Animal Facility except for the C57/Bl6 mice which were only used in the macrophage induction experiments and were purchased from Charles River UK Ltd. All animals were of conventional health status. Purchased animals were quarantined for 14 days prior to entry to the experimental animal room, animals bred at Ninewells were weaned at 21 days of age into plastic stock boxes and transferred to the experimental room 14 days later. The rats were housed in groups of 5 and the mice in groups of 10. All animals were fed on Special Diet Services, Rat and Mouse No. 1 maintenance diet. Tap water and diet were available ad libitum at all times. All stock animals had cages and bedding changed twice a week when they were handled during transfer to a clean cage.

2.2 Experimental environment

All experiments were carried out in the experimental animal room (3.5x6.0m). Except on rare occasions, eg visits of the Home Office Inspector and at the weekends when the on-duty staff watered and fed the animals, the room was never entered by anyone other than the first author, R J Barclay. Environmental conditions were maintained at $21 \pm 2^{o}C$ with 55 ± 10 per cent humidity, the room had 12 air changes per hour and was lit with artificial light; a 12:12h light:dark cycle, on at 0500h, was used.

The data recording equipment was housed in an ante-room, the door to which was kept closed during the experiments. The only variables in the room were therefore alterations in the air flow and associated sounds due to responses of the air conditioning system and the sounds of animal husbandry activities within the animal unit.

2.3 Housing

Rodents were caged in boxes made from high density polypropylene with stainless steel lids (North Kent Plastics). The dimensions of the boxes for rats were 50x30x25cm; while those for stock mice were 42x25x13cm and for experimental mice 30x13x13cm. Sterilized, autoclaved, whitewood shavings were used as bedding material. No nesting materials were supplied.

2.4 The activity monitor

Animal movements were detected by an electronic monitor. The method of operation of this instrument was such that it would be undetected by the subjects. In none of the experiments was any evidence found of feedback from the instrument influencing an animal's activities.

The monitor, a block diagram of which is shown in Figure 2, consisted of an amplified LC tuned circuit with a centre frequency just below that of the mains electricity supply. An LC circuit contains inductance and capacitance and resonates at a frequency dependent on the values of these components. In this case the capacitative element in the circuit was provided by an ordinary non-conductive plastic animal box with its rodent and other contents, the dielectric, and a conductive metal lid. The lid was connected to the rest of the monitor circuitry by a crocodile clip and flying lead. The electrical size of the capacitor varied with alterations in the dielectric caused by movements of the animal, a charged particle carrier.

Within the monitor circuitry the output of the tuned circuit was amplified, low-pass filtered and low-pass amplified before passing to a stage where the sensitivity of the circuit could be set by coarse and fine multi-turn potentiometers. The effect of the circuitry was to convert changes in frequency, caused by animal movements, to electrical pulses. The pulses were totalled by interfacing the monitor with a modified electronic calculator.

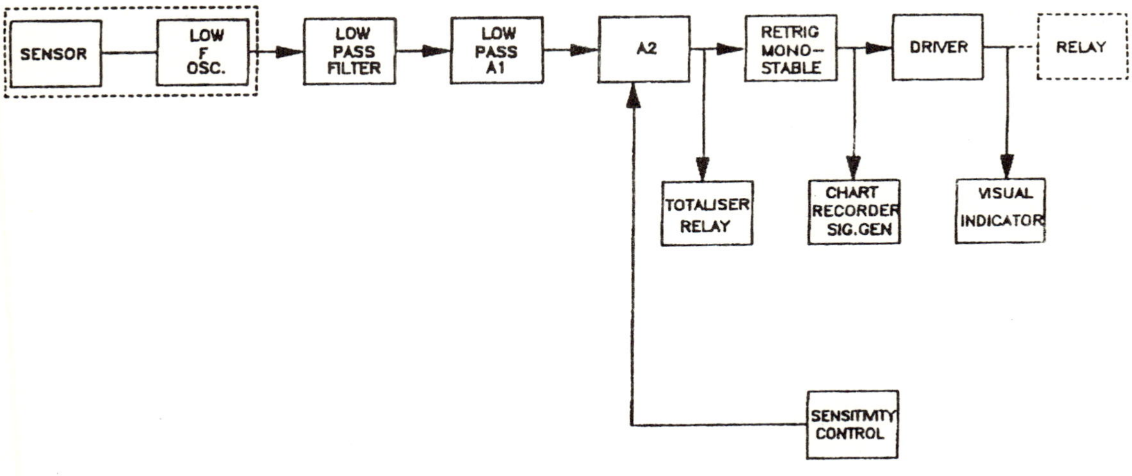

Figure 2 Diagramatic representation of the activity monitor.

Brief explanation: The SENSOR consists of the crocodile clip and metal cage lid, this detects LOW FREQUENCY OSCILLATIONS which are then LOW PASS FILTERED and amplified (LOW PASS A1). At the gain setting amplifier (A2) the sensitivity of the circuit is set. Data are output through the TOTALISER RELAY which operates a modified pocket calculator with printer. The RETRIGGERABLE MONOSTABLE returns the monitor to zero after a single event has occurred. The sensitivity level of the apparatus is displayed by the VISUAL INDICATOR. Optional features which were available (but not used) were the CHART RECORDER SIGNAL GENERATOR which could display an analogue signal and the RELAY which could be used to activate other equipment.

Formulae for calculating disturbance indices

	Period (minutes)	Disturbance Index
RAT	0-10	$DI^{10} = \log_{10} \frac{E^{10}}{C^{10}}$
	10-20	$DI^{20} = \log_{10} \frac{E^{20}}{C^{20}}$
	20-30	$DI^{30} = \log_{10} \frac{E^{30}}{C^{30}}$
	30-60	$DI^{60} = \log_{10} \frac{E^{40}+E^{50}+E^{60}}{C^{40}+C^{50}+C^{60}}$
MOUSE	0-30	$DI^{30} = \log_{10} \frac{E^{30}}{C^{30}}$
	30-90	$DI^{90} = \log_{10} \frac{E^{60}+E^{90}}{C^{60}+C^{90}}$
	90-210	$DI^{210} = \log_{10} \frac{E^{120}+E^{150}+E^{180}+E^{210}}{C^{120}+C^{150}+C^{180}+C^{210}}$

Key: E = experimental activity for the period
C = control activity for the period

Other types of movement detector would be suitable for use in generating a DI, providing that they did not selectively influence the behaviour of the animal.

The monitor was calibrated by placing an anaesthetized animal in the box and adjusting the sensitivity controls until respiratory movements did not trigger a movement record. As the animal recovered from the anaesthetic the sensitivity was further reduced so that whilst respiration and other non-locomotor activities such as body twitching were ignored, rearing and the movements associated with grooming were detected. Once calibrated, no changes were made to the sensitivity of the monitors during the course of the experiment.

2.5 Derivation of disturbance indices used for rats and mice

Rats and mice differ in the period that they spend in exploratory, marking and escape-seeking activities after introduction to a new environment. Rats carry out exploration for only a short period of time. They are very active for the first 20 minutes and this declines until all animals are resting after 60 minutes. Analysis of variance carried out on the records of their movements showed that this period can be divided into four segments, 0-10, 10-20, 20-30, and 30-60 minutes. Four indices were therefore calculated for rats, ie DI^{10}, DI^{20}, DI^{30} and DI^{60} where the superscript numbers indicate the end of each successive period of time from their introduction into the new environment.

In contrast, mice show a very high pattern of activity over the first 30 minutes. The pattern differs slightly between 30 - 90 and again between 90 - 210 minutes after which all animals rest. Three indices are, therefore, calculated for mice, ie DI^{30}, DI^{90} and DI^{210}. The DI^{30} was found to be particularly important in tests on mice because only exceptionally severe experimental procedures induced a change of activity in the first period. Usually the drive to explore and mark new territory was sufficiently imperative to overcome the effects of minor experimental procedures. The formulae used to calculate the indices for rats and mice are shown in the table opposite.

2.6 Testing the disturbance indices statistically

From the above formulae it will be apparent that in situations where the experimental results are identical to the control the fraction E/C will be unity and yield a DI of zero.

In each experiment, indices were obtained from ten animals for each period. The ten indices were then tested for their departure from zero using the Wilcoxon Matched-Pairs Signed-Ranks test described by Siegel (5).

2.7 Observation of exploratory activity

In pilot observations normal laboratory rats and mice were found to have habituated in their reactions to a novel environment after seven exposures. Subsequently their exploratory activity was highly predictable. The procedure used therefore, consisted of a series of habituation exposures, the tenth of which was scored, to provide the control figure.

All groups used in the test consisted of ten animals. The following procedures were employed for each observation. Rats were housed during each test in groups of five animals per box. They were taken from their home box at 0900h each morning for ten days and placed singly into a clean box for one hour. After this they were returned to their home box and box mates. On the tenth day their activity was recorded by a monitor attached to the box. On the eleventh day before placement into a clean box, the test procedure was performed on the animal. The rat was then placed into the box and its activity recorded to provide the experimental value. The same animal was paired with the same monitor for the control and test recordings. During the ten day habituation period the animal was also accustomed to the restraint or handling techniques required for the test procedure.

Mice were housed singly, in boxes of the same dimensions as those used for the test exposures, for the ten days prior to the procedure being performed. During this period the mice were transferred to fresh boxes daily at 0900h and were habituated to restraint and handling techniques. On day ten, a monitor was connected to each box at the time of transfer and the animal's activity recorded. On the eleventh day the test procedure was carried out, the animal placed in a fresh box to which a monitor had been attached and its activity again recorded. The same animal was paired with the same monitor for the control and test recordings. All activity monitoring commenced at 0900h.

2.8 Test procedures

Except where otherwise stated, all restraint procedures and injections were carried out by a person with extensive experience in handling the species used. The same person attended to all the animal husbandry needs of the stock animals such as feeding, watering and changing cages throughout the observational period. The animals were handled twice a week when placed in clean cages.

The routine injection and restraint techniques followed, in general, those described by Herbert and Kristensen (6).

For cell harvesting and counting procedures in the macrophage experiment, mice were anaesthetized with ether and then killed by cervical dislocation. The abdominal area was swabbed with 70 per cent alcohol before the skin incision was made. After incision the skin was deflected back to expose the abdominal muscle layer, this was lifted with sterile forceps to form a pocket. Three ml of sterile phosphate buffered saline (PBS) plus two ml of air were injected into the peritoneal cavity, the abdomen was massaged for one minute after which time one ml of fluid was removed by needle and syringe. The fluid was placed into a sterile tube and spun at 2,000 rpm for one minute, the supernatant was discarded and the cells resuspended in one ml of sterile PBS, this procedure was repeated a further three times.

After centrifugation the cells were resuspended in a 1 per cent solution of neutral red, made up with sterile PBS, and 0.025 ml of the suspension was pipetted onto a haemocytometer; macrophages were differentiated from other cells by the ingestion of the neutral red. Macrophages were counted in the four large corner squares of the haemocytometer. To determine if a significant increase in cell count was obtained, a Student's t test was performed comparing the number of cells harvested from a control group with the experimental agent group.

In the macrophage experiment, in addition to post-injection data, the DI was also assessed on the day of killing (pre-kill data=PK). The object of this measurement was to detect any persistent effect of the stimulating agent.

2.9 Presentation of results

In most instances the results of each observation are tabulated and also presented graphically.

In order to assist ready appreciation of the results, values for the DI are not given in the tables when the experimental movements did not differ significantly from those of controls; NS is then entered in the table. When the test results were found to differ from the controls at the $p < 0.05$ level using the Wilcoxon test the actual values for the DI are entered on the table. In a few cases where the probability lay between 5 and 10 per cent the value has been entered but marked with a plus sign (+).

The figures show the calculated DI; the points have been connected with straight lines to assist appreciation of trends of increased or decreased activity. When comparing figures for different experiments it should be noted that the vertical axes are not always on the same scale. The line at zero represents the expected value for the DI, ie the Null Hypothesis. The placement of the curve below or above this zero line indicates the deviation of the experimental group from normal but the slope of the curve also has important implications.

Where the DI falls below zero there has been a depression of activity but the direction of the slope indicates whether the fall is continuing or there is a compensating period of hyperactivity. Where the curve is above the zero line, an excess of activity has been recorded but the slope, once again, indicates the state of the animal's activity compared with the normal expected during that period of observation.

Figure 3 Disturbing effects on rats of the injection of: 'Solution A' (Haloperidol, 0.3mg/kg), 'Solution B' (Amphetamine 0.5mg.kg), 'Solution C' (0.9% NaCl).

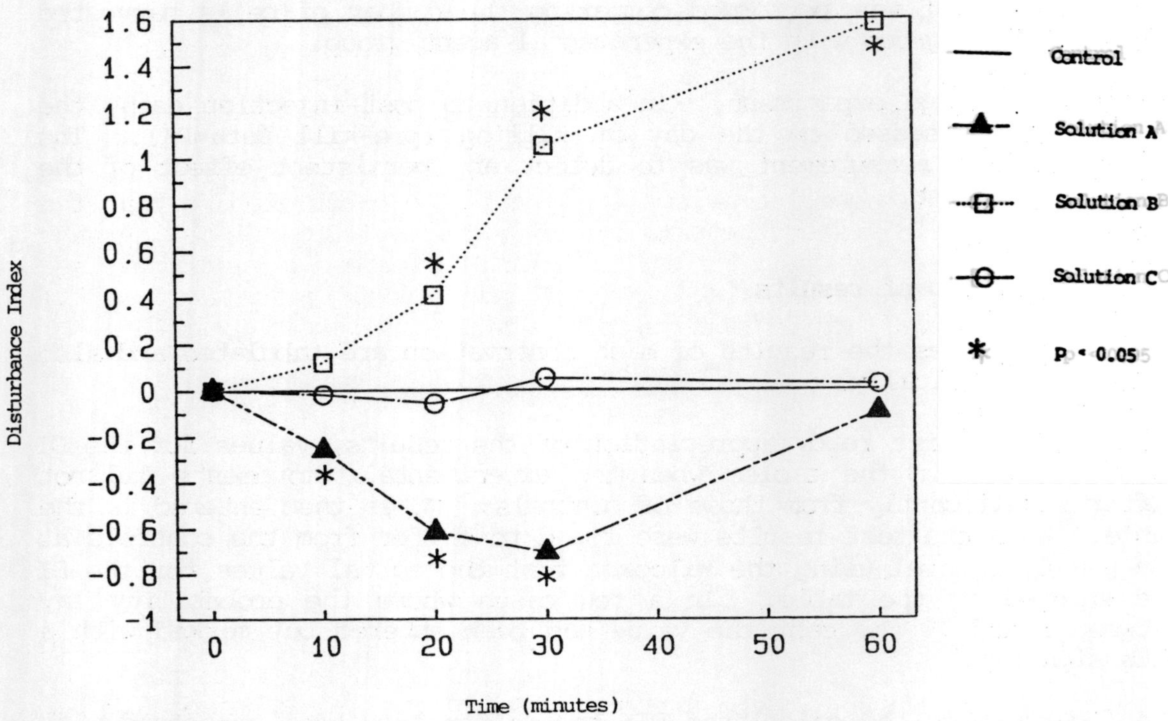

3. OBSERVATIONS ON RATS

3.1 Detection of the effects of drugs

As a general test of the validity of the DI, an experiment was designed to discover whether rats exposed to behaviour-altering drugs would yield indices that were consistent with the expected effect of the drugs.

In collaboration with Dr Balfour of the Department of Pharmacology, University of Dundee, a double blind procedure was designed for the test. Three substances were used:

Haloperidol, a tranquillizer which was expected to reduce activity;
Amphetamine sulphate, which was expected to increase activity;
As a control, 0.9 per cent saline, which should have no effect.

Solutions of these substances were prepared by a technician in the Department of Pharmacology and delivered to the animal facility in bottles marked A, B, and C. The concentrations of the drugs in the solutions were at therapeutic levels and had been calculated so that the volumes injected would be the same for each group of Sprague-Dawley rats tested. The observations were carried out as previously described, see Section 2, the solutions being injected subcutaneously just before the animals were placed into a new box for the activity record to be made.

The results were as follows and are shown graphically in Figure 3.

Solution	DI^{10}	DI^{20}	DI^{30}	DI^{60}
A	-0.28	-0.66	-0.73	NS
B	0.12	0.39	1.06	1.62
C	NS	NS	NS	NS

Dr Balfour examined these results and agreed that solution A appeared to have had a depressant effect, solution B a stimulant effect and solution C, no effect. The code was then broken and solution A found to be haloperidol, solution B to be amphetamine sulphate and C, to be 0.9 per cent saline. The indices therefore could be used to discriminate between the states of stimulation and depression in the rat.

3.2 Comparison of different means of manual restraint

Two simple means for the manual restraint of rats are commonly used in the laboratory. Both are well described in the literature. The first method, termed 'scruffing', is to grasp a fold of skin, the scruff, immediately behind the head of the animal. The second method is to place the hand over the animal's back and encircle the thorax with the thumb and fingers.

These two methods are easily applied by experienced handlers and readily accepted by laboratory rats. Inexperienced or frightened workers, on the other hand, often have difficulty in handling animals in these ways. They are hesitant, grasp the skin or thorax too tightly and show fear of being bitten. Because of this fear they are often tempted to wear gloves.

Figure 4 Variation in the Disturbance Index attributable to the method of restraining rats when held by an experienced handler.

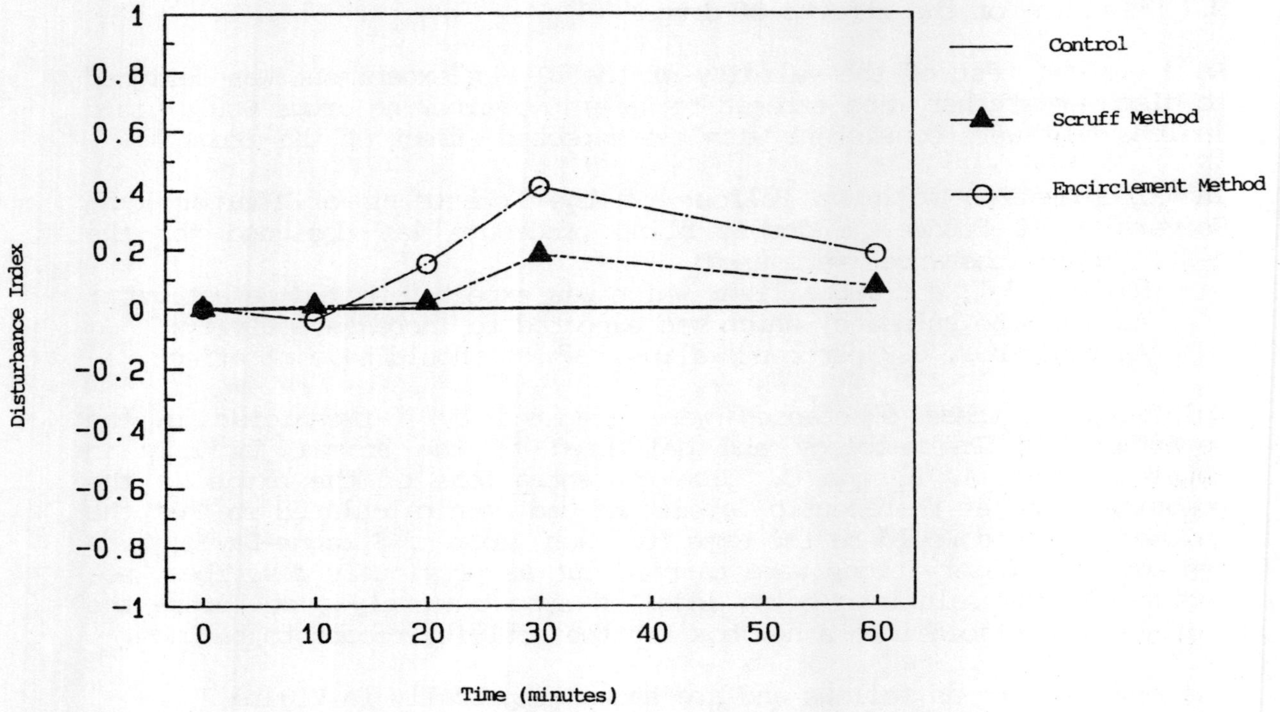

Figure 5 Variation in the Disturbance Index recorded for rats restrained for one minute by an inexperienced handler with or without thick gloves.

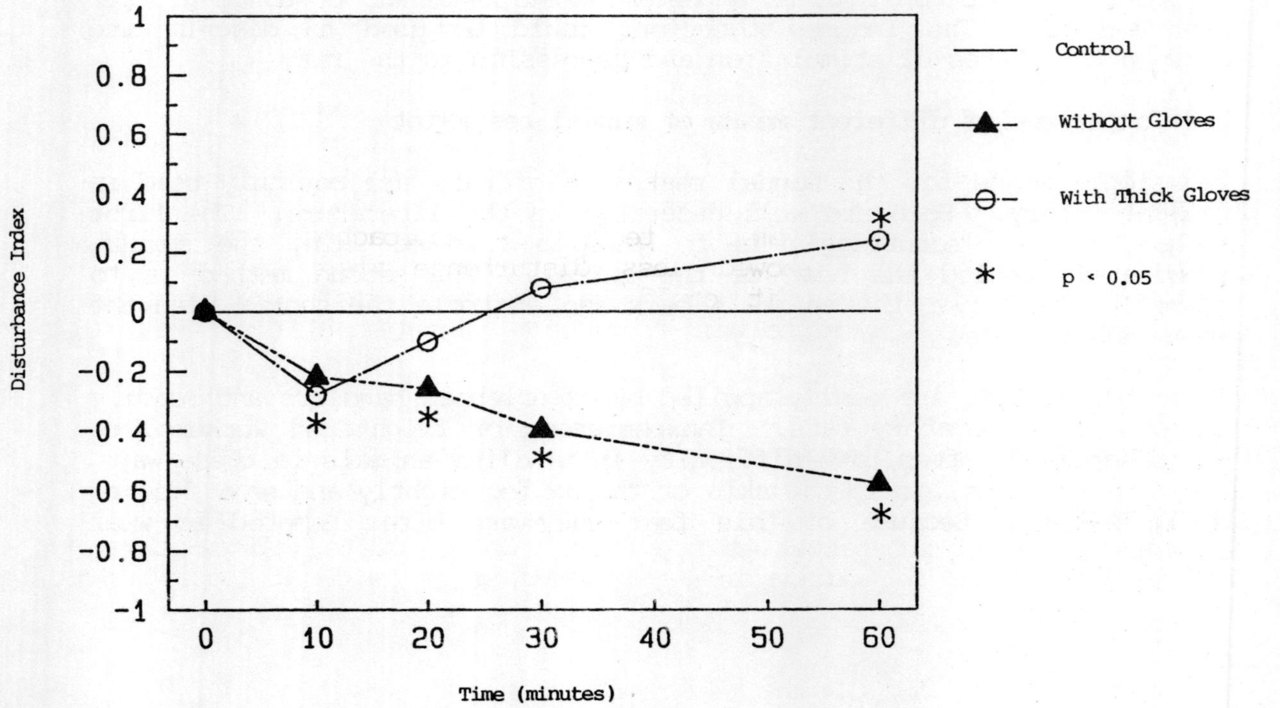

Tests were therefore carried out to find the method least disturbing when applied by an experienced handler and the disturbance, if any, produced by an inexperienced handler with and without gloves.

Both of the methods described for restraining rats, applied by an experienced handler, were compared for their disturbing effect. To decide which method should be tested with an inexperienced handler, forty senior honours biology undergraduate students, in an animal handling class, were questioned. They had no previous experience of handling rats and all expressed a preference for encircling the chest and would have used gloves, if available.

The inexperienced handler selected for the experiment was a recently recruited animal technician who had never previously handled rats, but was interested in and had some ability to handle other animals. He used the method of grasping the animal round the chest, with and without gloves. The gloves used were made of heavy duty asbestos.

In all cases the animal was removed from its home cage and restrained for ten seconds before being placed in the clean cage and its activity measured.

The results were as follows and are shown graphically in Figures 4 and 5.

	DI^{10}	DI^{20}	DI^{30}	DI^{60}
Experienced handler				
Scruff	NS	NS	NS	NS
Encirclement	NS	NS	0.41+	NS
Inexperienced handler using encirclement				
Without gloves	-0.22	-0.26	-0.40	-0.58
With gloves	-0.28	NS	NS	NS

+ $0.1 > p > 0.05$

It appeared that neither of the restraint techniques yielded a significant DI when applied by experienced staff, although the result at DI^{30} with the encirclement technique approached statistical significance. The rats showed less disturbance when inexperienced handlers wore gloves and it could be concluded that inexperienced handlers should be allowed to wear protective gloves should they wish to do so.

3.3 Intraperitoneal injections; influence of volume

Intraperitoneal injection is a frequently used route for the parenteral administration of substances to the rat. No objective information is however found in the literature, to serve as a guide to acceptable

Figure 6 Variation in the Disturbance Index for rats given 0.9% NaCl intraperitoneally at different dose rates.

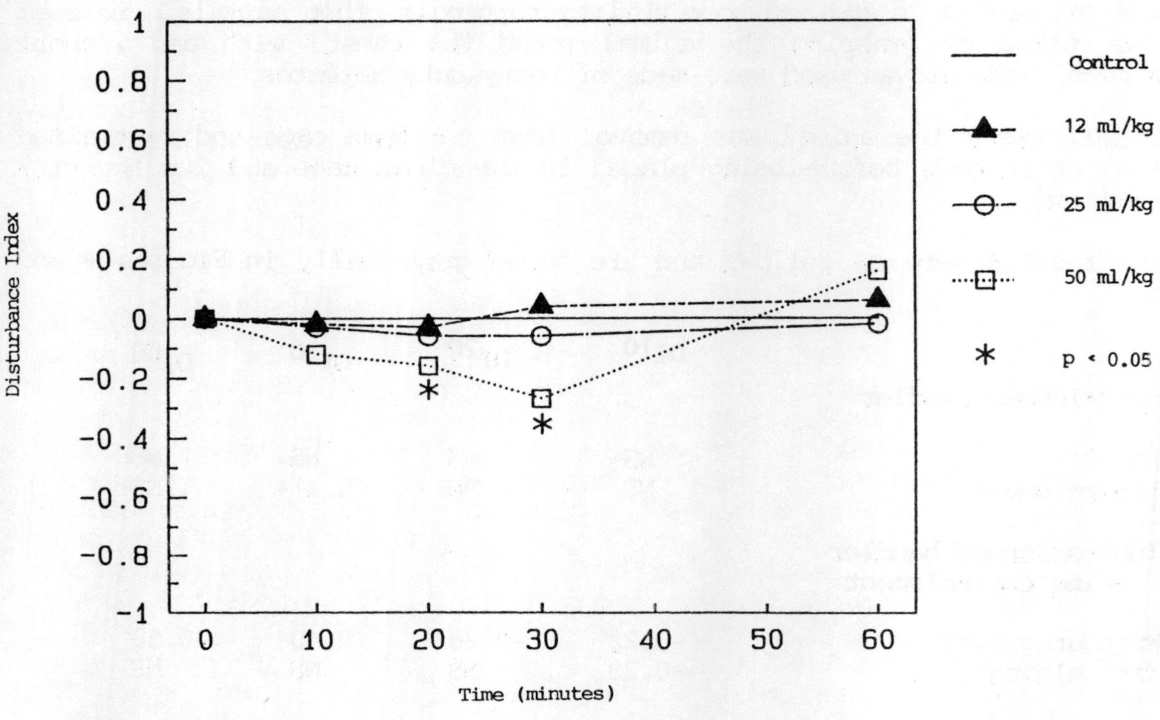

volumes for delivery by this route. Tests were therefore carried out to determine the effects on the DI of delivering various volumes of fluid to rats by the intraperitoneal route. The volumes given were the equivalents of injections of between 2.5 and 10 ml to a 200g rat.

The solution used was a 0.9 per cent solution of NaCl in distilled water at room temperature. The Sprague-Dawley rats, which had previously been weighed, were restrained by the handler and the injection made using a 0.5mm diameter needle at the volume/weight ratios shown. The record of exploratory activity was then made as follows:

Ratio	**DI10**	**DI20**	**DI30**	**DI60**
12ml/kg NaCl	NS	NS	NS	NS
25ml/kg NaCl	NS	NS	NS	NS
50ml/kg NaCl	NS	-0.16	-0.27	NS

The variations in the DI relating to volumes of inoculum are shown in Figure 6. Only at the volume/weight ratio of 50ml/kg were there any significant changes. These represented a minor diminution, lasting for about 20 minutes, followed by a rebound in activity.

3.4 Foot pad injection

The foot pad is frequently used as an injection site in immunological experiments. It is believed that absorption is slow from this site and antigens placed there have a maximum opportunity to stimulate a response, especially in the popliteal lymph node. Freund's complete adjuvant is often used to augment the effect, though it usually produces necrosis of the foot. Foot pad injection is also used in tests for hypersensitivity. The thickness of the foot is measured before and at intervals after injection of the test substance. The swelling which results is probably painful. Even the insertion of a needle into this site without injection produces signs of pain and fear as the foot is pricked. Though it is evident that this site should not be used, observations were made so as to yield indices for such injections. The object was that these indices could later be compared with those for other procedures as a guide to the severity of injections given by this route.

A sterile solution of 0.9 per cent NaCl in water was used for injections into one hind foot of a Sprague-Dawley rat. The dose rate of 0.15ml/kg is the equivalent of 0.03ml in a 200g rat. In practice, larger rats were used with an injection volume of 0.05ml which is the smallest volume that can conveniently be measured using a plastic 1ml syringe. A recording of exploratory activity was made immediately after the injection and thereafter the rat was returned to its home box. Twenty-four hours later another recording of activity was made to detect any residual effects of the injection.

Figure 7 Disturbance produced immediately and twenty-four hours after rats had been injected with 0.9% saline into one hind foot pad.

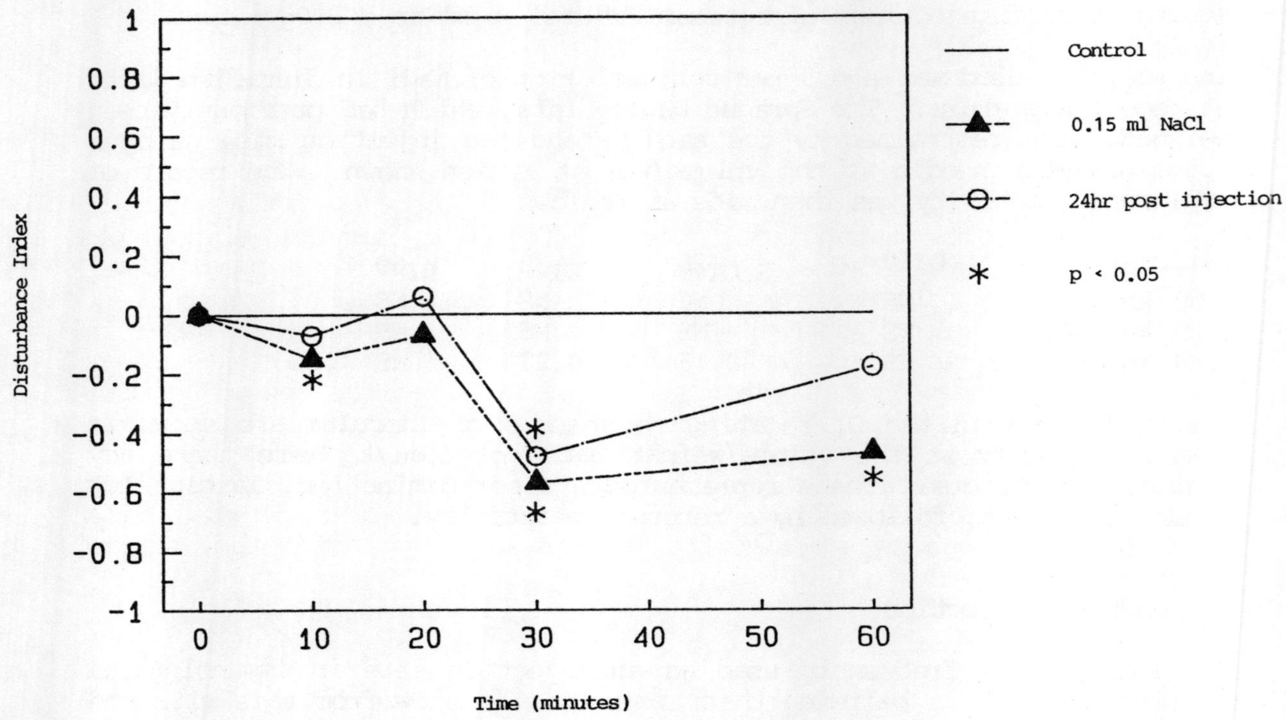

Figure 8 Disturbance of rats caused by the administration of tap water <u>per os</u> (by cannula).

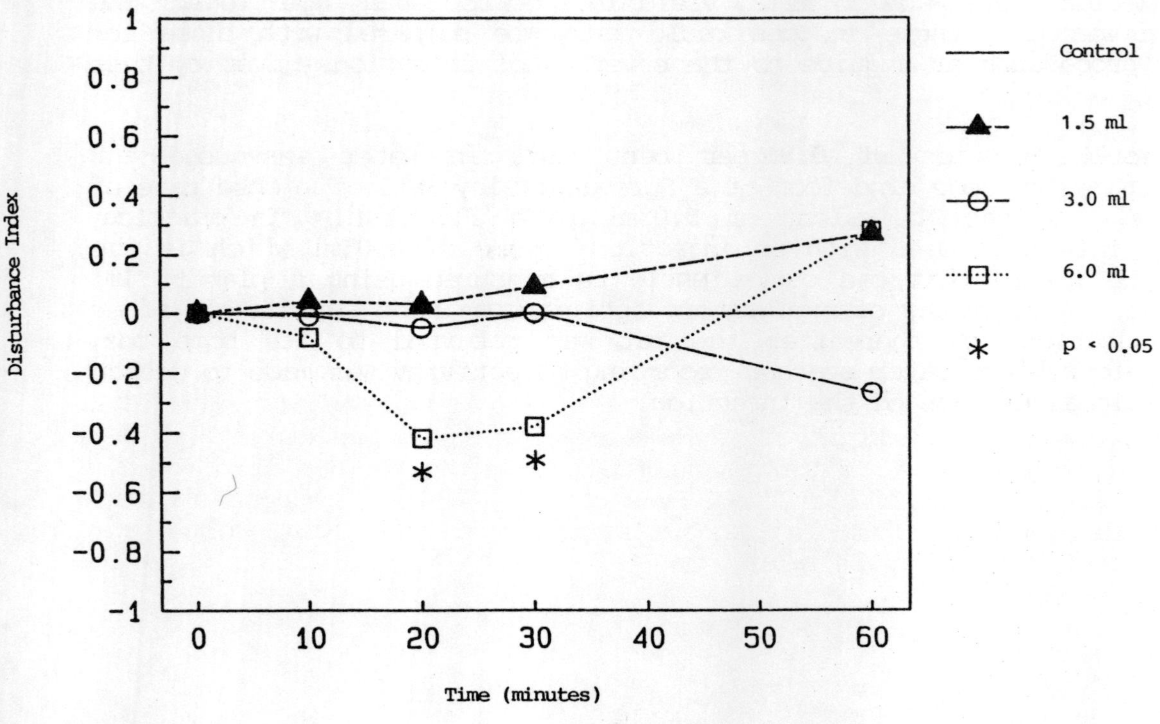

The results, shown graphically in Figure 7, were as follows:

Timing	DI^{10}	DI^{20}	DI^{30}	DI^{60}
Immediately post-injection	-0.15	NS	-0.57	-0.47
24h post-injection	NS	NS	-0.48	-0.18+

+ $0.1 > p > 0.05$

These results show that this route of injection not only causes an immediate high level of disturbance but that the disturbance persists at least until the following day. The solution used was bland and no overt inflammation was seen, but in many experiments in which this route is used this is not the case, so that the effects must be much more acutely disturbing to the animals. This experiment yielded the second highest DI(-0.57) in this Report apart from those from macrophage induction in mice. The extent to which the decrease in exploration was a direct result of pain in the foot pad cannot be determined. However this result clearly indicates that the foot pad should never be used as a site for injections and as alternative sites are available, they should be used.

3.5 Administration of substances by mouth

Placement of substances directly into the stomach by oesophageal cannulation is a common procedure and is used extensively in the pharmaceutical field to test drugs. Tests were carried out to examine the disturbing effect, if any, of commonly administered volumes of liquid by this route.

Plain tap water, at room temperature, was used as the liquid to be given. No preliminary treatment, such as starvation, was applied to the Sprague-Dawley rats which were taken directly from their home boxes, dosed, and their exploratory activity recorded. The water was administered through a standard, curved, metal cannula, by an experienced handler fully conversant with the method. The largest dose given was the equivalent of 1.2ml for a 200g rat.

The results were as follows and are shown graphically in Figure 8.

Volume	DI^{10}	DI^{20}	DI^{30}	DI^{60}
1.5ml/kg	NS	NS	NS	NS
3.0ml/kg	NS	NS	NS	-0.27+
6.0ml/kg	NS	-0.42	-0.37	0.28+

At the 6.0ml/kg rate the disturbance caused between 10-30 minutes after administration is profound, though it is followed by a rebound of rapidly increasing activity. That such disturbance is caused by what would appear to be a relatively moderate dose, indicates that ideally volumes should be much less. The rate of 6.0ml/kg must be assumed to be an absolute upper limit for innocuous substances that can probably be tolerated, as recovery is rapid.

3.6 Summary of observations on rats

The results of the experiments are re-presented here for comparative purposes and indicate the wide applicability of the DI method for assessing the relative severity of procedures on rats.

EXPERIMENT	DI^{10}	DI^{20}	DI^{30}	DI^{60}
BLIND EXPERIMENT				
0.3mg/kg Haloperidol	-0.28	-0.66	-0.73	NS
0.5mg/kg Amphetamine sulphate	0.12	0.39	1.06	1.62
0.9% Saline	NS	NS	NS	NS
MANUAL RESTRAINT				
Experienced				
- Scruff	NS	NS	NS	NS
- Encirclement	NS	NS	0.41+	NS
Inexperienced				
- Without gloves	-0.22	-0.26	-0.40	-0.58
- With gloves	-0.28	NS	NS	NS
EFFECT OF VOLUME				
12ml/kg NaCl i.p	NS	NS	NS	NS
25ml/kg NaCl i.p	NS	NS	NS	NS
50ml/kg NaCl i.p	NS	-0.16	-0.27	NS
INJECTION INTO FOOT PAD				
Immediately post-injection	-0.15	NS	-0.57	-0.47
24 h post-injection	NS	NS	-0.48	-0.18+
ADMINISTRATION PER OS				
1.5ml/kg	NS	NS	NS	NS
3.0ml/kg	NS	NS	NS	-0.27+
6.0ml/kg	NS	-0.42	-0.37	0.28+

+ $0.1 > p > 0.05$

4. OBSERVATIONS ON MICE

4.1 Duration of restraint

Restraint times for inoculations vary considerably and are influenced by the experience of the handler, the size of needle being used and the viscosity of the solution being injected. Some strains of mice accept restraint by an experienced handler for some seconds before they struggle and try to escape, others object vigorously to restraint of any description.

Three durations of restraint, within a range likely to apply in practice, were examined to discover whether they caused disturbance to the exploratory behaviour of Swiss mice.

On the eleventh day of habituation, and in the process of transfer to the new box for recording, Swiss mice were picked out of their home box by the tail by an experienced handler, placed onto the cage lid and grasped by the scruff of the neck. They were then held in one position on the bench for measured periods of 10, 20 or 30 seconds before being placed in the new box for recording.

The results were as follows and are also shown graphically in Figure 9.

Period	DI^{30}	DI^{90}	DI^{210}
10 seconds	NS	NS	NS
20 seconds	-0.16	NS	NS
30 seconds	-0.20	-0.18	0.14

Figure 9 Disturbing effects on mice of different durations of restraint.

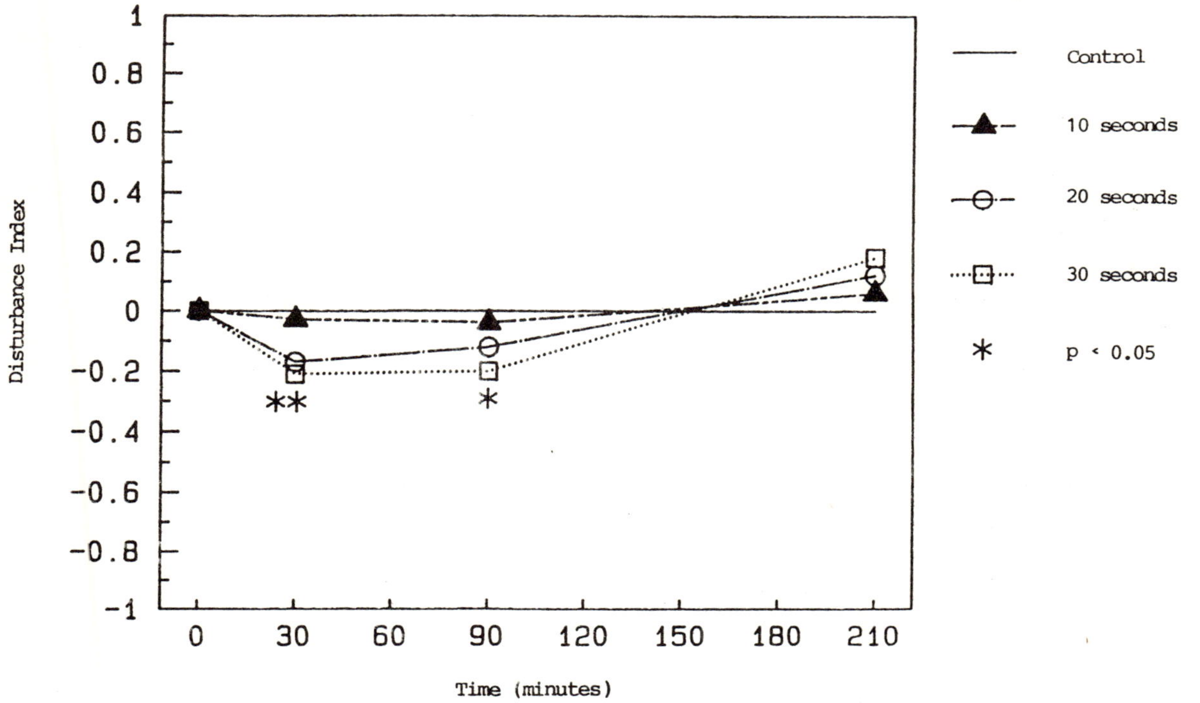

The results indicate that mice should be manually restrained as briefly as possible, the longer they are restrained the greater the DI. The positive index at DI^{210} for the 30 second restraint may be a rebound phenomenon resulting from previous underactivity.

4.2 Unskilled handling and scent contamination

Persons carrying out experiments with mice gain skill and confidence with time. It has already been shown that inexperienced persons handling rats cause significant disturbance, see Section 3.2. The difficulty of controlling relatively large animals and the mutual fear engendered between them and humans makes this understandable. Mice, on the other hand, are more easily controlled and a test was designed to discover whether any disturbance is caused by inexperienced handling.

During the habituation period Swiss mice were transferred to the recording cage by the usual, experienced handler holding the base of the tail. On day eleven the transfer was done by an inexperienced person. The animal was not held or restrained, except momentarily, during the transfer.

The inexperienced handlers, technical assistants within the unit who did not handle animals, had received a minimum of instruction including a demonstration of the method of transferring animals from cage to cage; there was therefore, no question of an unacceptable technique being employed. The transferred animals were manipulated by the tail only.

Figure 10 Disturbance caused to mice by transfer from box to box by an inexperienced handler, and by an experienced handler, contaminated with the urine of a domestic cat.

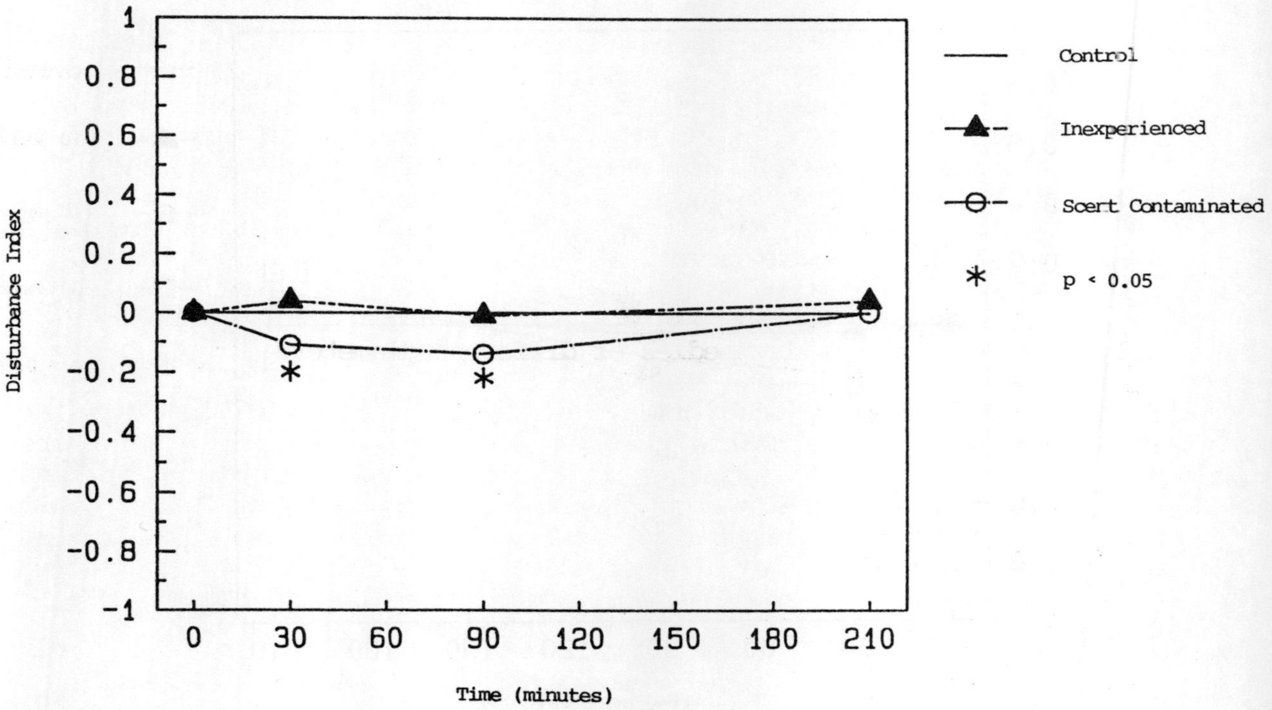

Another problem, an under-appreciated one, arises from the variety of species kept within animal facilities. In the wild, some of these have a predator/prey relationship. In well organized animal facilities different members of the staff attend to these different species or care for them in a suitable sequence. Experimenters however, are unlikely to be aware of such nuances and move directly from predator to prey carrying the scent of the predator with them. A test was therefore devised to determine whether or not such contamination would be detected by the prey species to the extent that it had a disturbing effect on them.

On the eleventh day of habituation an experienced handler contaminated his hand with the urine of a spayed female domestic cat. This was mopped onto the skin and allowed to dry so that the mice handled did not become contaminated with wet urine. The mice were transferred to the recording box using this hand, each animal being picked up and placed onto the lid of the cage and stroked along its back. The transfer was direct and quick with only momentary restraint.

The results of these two tests were as follows and are shown graphically in Figure 10.

	DI^{30}	DI^{90}	DI^{210}
Inexperienced handler	NS	NS	NS
Experienced, Scent contaminated handler	-0.11	-0.14	NS

Transfer from box to box by an inexperienced handler produced no significant disturbance despite the fact that the animals became agitated and sought to escape when being caught in their home box.

Contamination of the transferring hand with cat urine had a small, but significant, effect. This is probably the first time that such an effect has been objectively detected and measured. The disturbance took the form of a diminution of activity during the first 90 minutes and it is notable that it was not followed by a rebound of exploration of the novel environment. This suggests that the effect was very potent, confining the animal to the minimum activity necessary to establish a knowledge of its surroundings.

4.3 The use of hypodermic needles of different diameters

The hypodermic needles most often used for injections in mice are the smallest diameter disposable needles commonly available, which are 0.5mm. However when a viscous solution or one containing particles has to be injected, larger diameter needles, 0.63mm, 0.8mm or 1.1mm, must then be used and these might have some effect on the mice. If the head and trunk length of a human is equated with that of a mouse then the use of a 1.1mm needle in a mouse is equivalent to the use of 1.1cm needle in a human.

Tests were designed to investigate whether the use of any of the commonly employed diameters of hypodermic needle had a disturbing effect on mice. The tests were carried out by holding the animals, 30g Swiss mice, as for intraperitoneal injection and passing the needle through the skin and body wall. No material was injected, the needle was at once withdrawn and the mouse placed in a clean box for the activity recording to be made.

The results were as follows:

Needle size	DI^{30}	DI^{90}	DI^{210}
1.1 x 40mm	NS	NS	NS
0.63 x 28mm	NS	NS	NS
0.5 x 25mm	NS	NS	NS

None of the needles used produced any significant disturbance. Despite the relatively large size of the needles by human standards this could be because, if they are to survive in the wild, animals cannot afford to be even slightly incapacitated by insignificant traumas. It is not then surprising that needle pricks have no significant disturbing effect. Where there is only minor tissue damage any effect is presumably confined to the moment of trauma alone.

4.4 Differing routes of injection

The most frequently used routes for the parenteral administration of solutions to the mouse are intraperitoneal (i.p.) and subcutaneous (s.c.). The intramuscular (i.m.) route is also occasionally employed, but the intravenous much more rarely, due to the time required and its technical difficulty. In many instances the route is determined by the requirements of the experiment, while in others there is a choice. Tests were carried out with the object of determining whether, when a choice is possible, one route should be chosen in preference to another, because it has a less disturbing effect.

The solution used for the injections in Swiss mice, was 0.9 per cent sterile saline. It was delivered using a 0.5mm hypodermic needle at a dose rate of 12ml/kg for the s.c. and i.p. routes and at half that for the i.m. route. At 12ml/kg the dose for a 30g mouse is 0.36ml, which was considered unacceptably large to give i.m., hence the reduced dosage. In tests of i.p. injections, recorded in this Report, see Section 4.5, a rate of 12ml/kg of 0.9 per cent NaCl has been found not to cause disturbance.

The results were:

Volume/rate	DI^{30}	DI^{90}	DI^{210}
12ml/kg i.p.	NS	NS	NS
12ml/kg s.c.	NS	NS	NS
6ml/kg i.m.	NS	NS	NS

At the dosage rates employed and the routes used, the injection of 0.9 per cent NaCl caused no significant disturbance to mice.

4.5 Volume and substrate of intraperitoneal injections

Intraperitoneal (i.p.) injection is probably the most commonly used route for the administration of substances to the mouse. It can easily and rapidly be carried out by one person, the skill required is minimal, relatively large volumes can be given and absorption of substances from this site is rapid as there is direct drainage from the peritoneal cavity to the vena cava via the thoracic duct. Despite the importance of the route, there appear to be no recommendations, based on objective evidence, as to the volume that it is advisable to give i.p. Volumes of 2 or 3ml are often recommended; a volume of 3ml in a 20g mouse is equivalent to 10.5l (2.3 gallons) in a 70kg (11 stone) man.

The effect of the administration of a large volume of fluid by the i.p. route is both mechanical and physiological. Mechanically there can be hindrance to movement due to the sudden increase in abdominal size together with pain from irritation of the peritoneum and at the injection site. Possibly more important are the physiological changes brought about by the material injected. The water would lead to increased urination but the solutes may grossly disturb the basic biochemical equilibrium of the animal and this effect may persist for a long time.

Tests were carried out to determine the immediate effects on the DI of delivering various volumes per weight unit. The maximum being the equivalent of injecting 1.5ml by this route into a 30g mouse. Two solutions were used: a 0.9 per cent solution of NaCl in distilled water and Hartmann's balanced salts solution. The NaCl solution is the 'normal saline' extensively used for all manner of purposes when dealing with animals, their organs, and their tissues. The balanced salts solution is a saline solution that is more physiologically complete and compatible with mammalian tissues and fluids. Both solutions were used at room temperature. The Swiss mice, which had previously been weighed, were restrained by the handler and the injection made using a 0.5mm diameter needle at the volume/weight ratios shown. The record of exploratory activity was then made.

The results which are shown graphically in Figures 11 and 12 were as follows:

Volume	**DI^{30}**	**DI^{90}**	**DI^{210}**
12ml/kg NaCl	NS	NS	NS
25ml/kg NaCl	-0.07	-0.13	NS
50ml/kg NaCl	-0.28	-0.53	-0.34
50ml/kg Hartmann's	NS	NS	NS

Figure 11 Disturbance of mice produced by the intraperitoneal injection of different volumes of saline (0.9% Nacl).

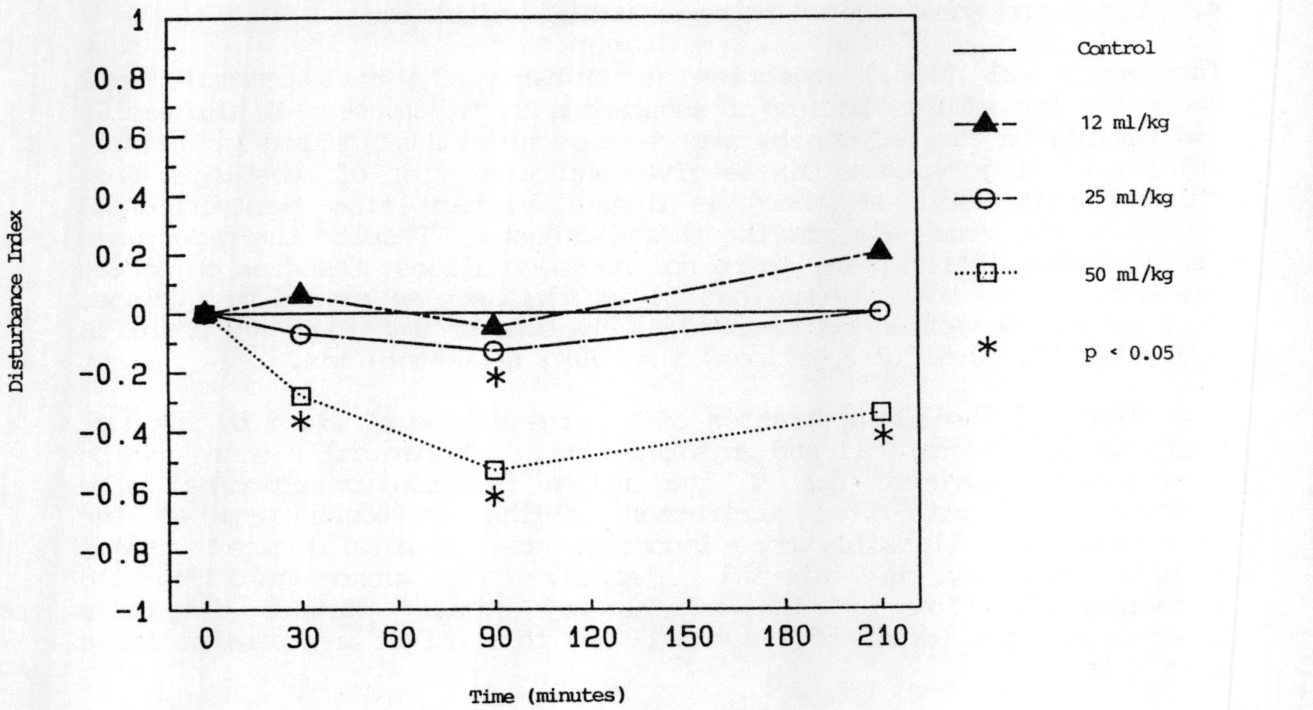

Figure 12 Disturbance of mice produced by the intraperitoneal injection of saline (0.9% NaCl) and Hartmann's solution at 50ml/kg (1.5ml for a 30gm mouse).

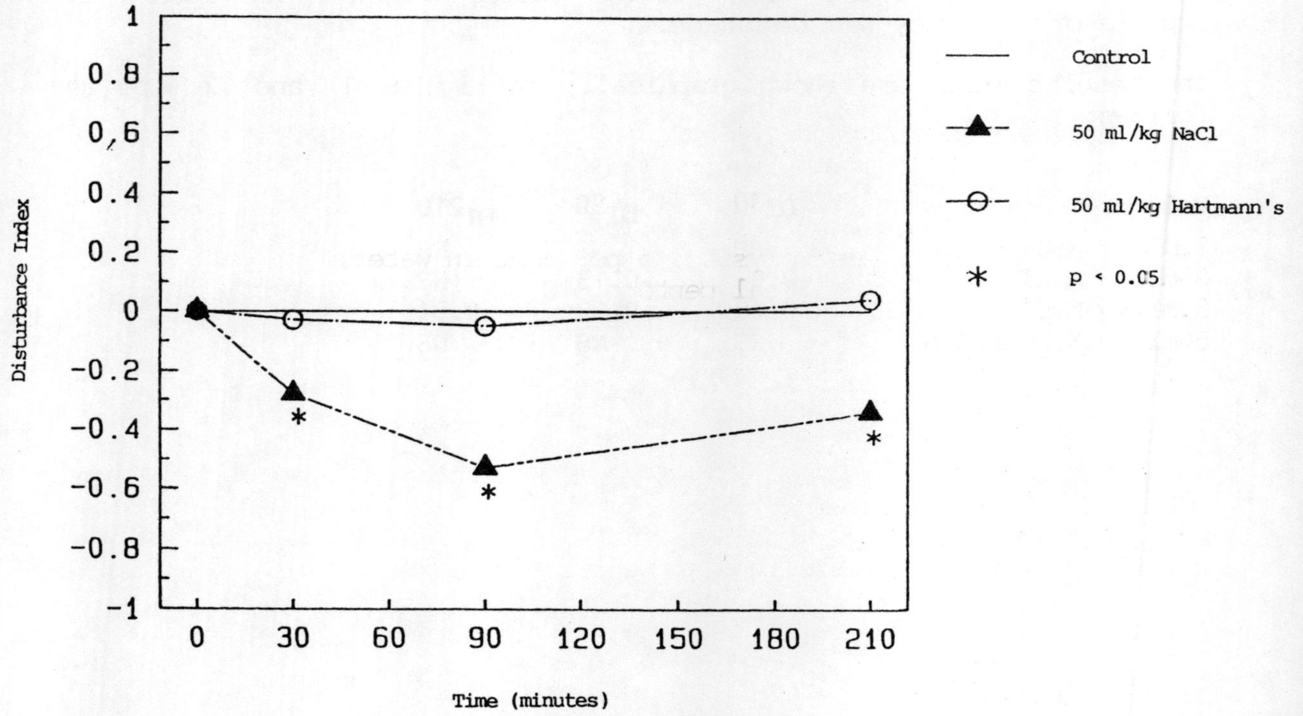

The three volume/weight ratios used correspond to volumes of 0.36, 0.75 and 1.5ml for a 30g mouse. The indices recorded for the 25ml/kg rate are significantly different from the controls but are small and probably indicate the largest volume of this solute that it is acceptable to deliver i.p. to a mouse. At twice this volume/weight ratio (50ml/kg) the DI is large and unacceptable. If, however, a balanced salts solution is substituted for the 0.9 per cent NaCl a dramatic change is seen in that there is then no significant disturbance of the animal's normal exploratory activity.

A further observation of interest is the species difference seen in the responses of rats and mice to a volume/weight ratio of 50ml/kg of the 0.9per cent NaCl. Whilst this ratio is disturbing to the mouse it is tolerated by the rat, see Section 3.3. It is possible therefore, that even larger volumes could be given to rats if a balanced salts solution was used. The difference between rats and mice also emphasizes that the responses of one species may not be a good guide to those of another.

4.6 Peritoneal macrophage induction: the effects of inducing agents

Macrophages are tissue and body cavity resident phagocytic cells which have somewhat similar functions to the phagocytic white cells of the blood. Additionally however, they are important intermediates in the initiation of immune responses and are therefore much studied. They are most commonly obtained from the peritoneal cavity by lavage and the yield is greatly increased if an inducing substance has been injected into the cavity a few days before harvesting the cells. Many of these substances are irritant and mice show distress after they have been injected.

Nine commonly employed inducing substances were examined with the aim of finding those with the least disturbing effect. As the least disturbing might also be the least useful for the purpose intended, an estimate of the relative macrophage yield induced by each substance was also made. The doses administered were those in common practice and no effort was made to standardize for injected volume. Any disturbing effects would therefore depend both on the nature of the substance injected and its volume.

The nine inducing substances tested were:

- Calf serum.
- Proteose peptone broth, 10 per cent.
- Glycogen, 5 per cent in water.
- Casein hydrolysate, 5 per cent in water.
- Bacteriological peptone, 10 per cent in water.
- Starch, 1 per cent in 0.9 per cent NaCl.
- Mineral oil (light liquid paraffin B.P.).
- Glycerol tricaprate.
- Thioglycollate broth, 3 per cent.

The bacteriological media, peptone water and thioglycollate broth, were prepared after the recipes given by Cruickshank (7). The proteose peptone broth was made up from dried medium (Oxoid). All of the substances used were sterilized appropriately before use.

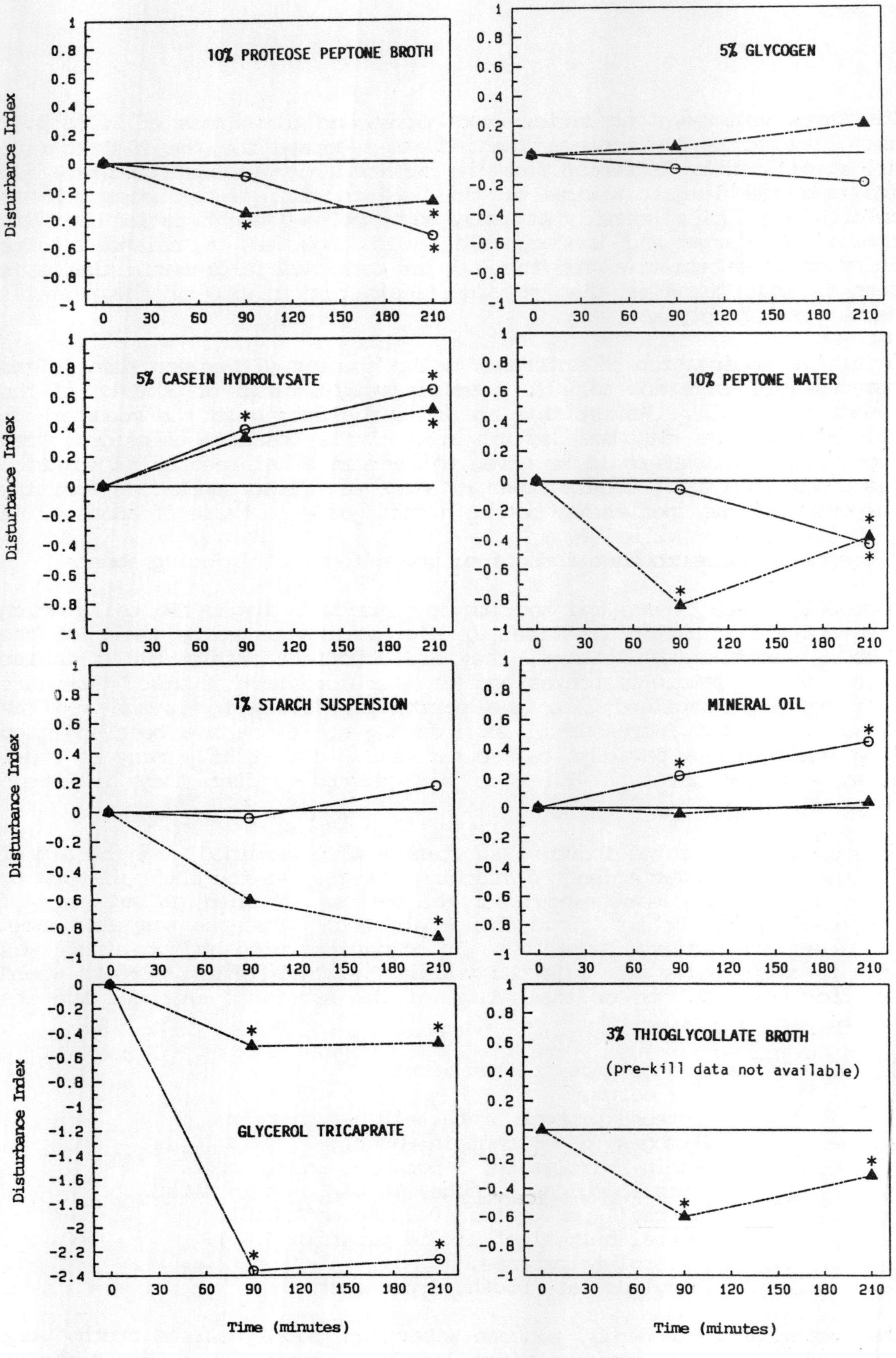

Figure 13 Macrophage induction: disturbance of mice immediately following intraperitoneal injection and pre-kill data for a variety of inducing substances.

The tests were all conducted with male C57/Bl6 inbred mice. Two sets of indices were obtained for each substance. The first post-injection or PI, was calculated from a recording of activity made immediately after the material had been injected. The second, pre-kill or PK, was calculated from exploratory activity recorded just before the animal was killed to harvest the macrophages for counting. The interval allowed for macrophage accumulation was that used traditionally and varied from 3 to 6 days as shown in the table of results.

Two indices were obtained as follows: on day eleven of habituation each animal was injected with the inducing agent by an experienced experimenter using a 0.5mm diameter needle. It was then placed immediately into a clean box and the exploratory activity monitored. This supplied the data for the PI index. After passage of the interval chosen for macrophage accumulation, the animal was again placed in a clean box and its activity monitored. This second activity recording provided the data for the PK index. At the end of the second monitoring the animal was killed and a peritoneal macrophage count made as described in Section 2.8 of this Report.

The results were as follows and are shown graphically in Figure 13.

INDUCING AGENT	**DOSE** (ml)	**INTERVAL** (days)	**CELLS/** ml(10^6)	**POST-INJECTION** (PI) DI^{90}	DI^{210}	**PRE-KILLING** (PK) DI^{90}	DI^{210}
Calf serum	0.5	3	0.4	NS	NS	NS	NS
10% Proteose peptone broth	1.5	5	0.4	-0.36	-0.28	NS	-0.52
5% Glycogen	0.5	5	0.48*	NS	NS	NS	NS
5% Casein hydrolysate	0.2	4	0.5**	0.32	0.51	0.38	0.65
10% Peptone water	1.0	5	0.7**	-0.84	-0.38	NS	-0.43
1% Starch in saline	2.0	3	1.3**	-0.61	-0.88	NS	NS
Mineral oil	0.5	6	1.8**	NS	NS	0.22	0.45
Glycerol tricaprate	2.0	5	#	-0.51	-0.49	-2.36	-2.27
3% Thioglycollate	3.0	6	##	0.58	-0.31	80% MORTALITY	
None	-	-	0.36	-	-	-	-

Key * cell count significantly different from control $p<0.001$ (Student's t test)
** cell count significantly different from control $p<0.0001$ (Student's t test)
# mice very near death 5 days post-injection; few single cells obtained from 80 per cent of the animals. The majority of cells were in large clumps making counting impossible.
eight of the ten animals died before the sixth day. Owing to the high mortality this experiment was not repeated.

Note DI^{90} represents the period 0-90 minutes. DI^{30} not calculated.

Three of the macrophage inducing substances produced no immediate disturbing effect indicating they were bland. With two of these, calf serum and glycogen, there was no long term effect either but the increase in yield of macrophages, though statistically significant in the case of glycogen, was minimal. The third, mineral oil, produced a five-fold increase in macrophage yield and this was greater than that produced by any other substance tested. At the time of harvest however, considerable disturbance was recorded, taking the form of an increase in activity. If the mice had been suffering from abdominal pain they would be expected to show a reluctance to move. The etiology of the increased activity seen with both mineral oil and casein is unknown.

The next most active material was starch. This produced no long lasting disturbance and presumably no damage to the peritoneum, pain, or irritation. It was however, very disturbing on injection. Some of this disturbance may however be related to the volume given and the nature of the vehicle used. Earlier in this Report experiments are described in which a volume of 1.5ml of saline injected i.p. into a mouse produced a rather similar but less marked pattern of disturbance; in the present experiment 2.0ml was used. A reduction in volume should reduce the disturbance encountered. Even more usefully the substitution of a balanced salts solution might eliminate it entirely, see Section 4.5.

The other substances tested were highly disturbing or actually toxic. None of them showed any advantages in macrophage induction and there seems to be no reason for their continued use.

Induction of peritoneal macrophages within the peritoneal cavity probably results from the provision of particles or debris within the cavity to which the macrophages are attracted. Inducers provide the particles, either by producing a peritonitis, or because they themselves are particulate. The ideal inducer may therefore, consist of non-irritant particles. From the results of the indices starch is disturbing only at injection and this may be a volume effect. Mineral oil is bland and becomes emulsified within the cavity to provide the particles. Some forms of oil are however, known to be irritant to the peritoneum and this property has been extensively investigated in connection with the use of oil as a vehicle for vaccines. Mineral oils, tested for irritancy (8), can now be obtained but their efficiency in inducing macrophages is unrecorded. The light liquid paraffin used here is known to be irritant in this context.

Tests for disturbance and irritancy could provide the tools to search for the most humane and effective means for inducing macrophages in mice and, by extension, in other animals. Until this is done starch appears to be the best macrophage inducer to use in preference to all other substances. Mineral oil has classically been recommended for the induction of macrophages and, from these results, it may still be acceptable for some purposes. If these results are considered from the viewpoint of the sum total of suffering inflicted on mice, it would appear to be more humane to avoid the use of macrophage inducing substances altogether and simply to euthanase undisturbed mice. This would involve using more animals but the undisturbed mice will not have been subjected to a distressing experiment and will not have suffered.

4.7 Summary table of observations on mice

As in the case of the experiments in rats, the DI method may be used to compare the severity of a wide range of procedures in mice. It is also evident that there are marked differences in response between the species.

EXPERIMENT	DI^{30}	DI^{90}	DI^{210}
EFFECT OF RESTRAINT TIME			
10 seconds	NS	NS	NS
20 seconds	-0.16	NS	NS
30 seconds	-0.20	-0.18	0.14
HANDLING EXPERIMENT			
Cat scent handler	-0.11	-0.14	NS
Inexperienced handler	NS	NS	NS
EFFECT OF NEEDLE SIZE			
1.1 x 40mm	NS	NS	NS
0.63 x 28mm	NS	NS	NS
0.5 x 25mm	NS	NS	NS
EFFECT OF ROUTE OF INJECTION			
12ml/kg NaCl i.p.	NS	NS	NS
12ml/kg NaCl s.c.	NS	NS	NS
6ml/kg NaCl i.m.	NS	NS	NS
EFFECT OF VOLUME			
12ml/kg NaCl i.p.	NS	NS	NS
25ml/kg NaCl i.p.	-0.07	-0.13	NS
50ml/kg NaCl i.p.	-0.28	-0.53	-0.34
50ml/kg Hartmann's i.p.	NS	NS	NS

MACROPHAGE INDUCTION

INDUCING AGENT	DOSE (ml)	INTERVAL (days)	CELLS/ ml(10^6)	POST-INJECTION (PI)		PRE-KILLING (PK)	
				DI^{90}	DI^{210}	DI^{90}	DI^{210}
Calf serum	0.5	3	0.4	NS	NS	NS	NS
10% Proteose peptone broth	1.5	5	0.4	-0.36	-0.28	NS	-0.52
5% Glycogen	0.5	5	0.48*	NS	NS	NS	NS
5% Casein hydrolysate	0.2	4	0.5**	0.32	0.51	0.38	0.65
10% Peptone water	1.0	5	0.7**	-0.84	-0.38	NS	-0.43
1% Starch in saline	2.0	3	1.3**	-0.61	-0.88	NS	NS
Mineral oil	0.5	6	1.8**	NS	NS	0.22	0.45
Glycerol tricaprate	2.0	5	#	-0.51	-0.49	-2.36	-2.27
3% Thioglycollate	3.0	6	##	0.58	-0.31	80% MORTALITY	
None	-	-	0.36	-	-	-	-

5. DISCUSSION

The observations show that by monitoring changes in the normal exploratory behaviour of an animal that has been placed in a new environment, a Disturbance Index or DI can be obtained. This index is sensitive to changes in exploratory behaviour caused by handling, restraint, or experimental procedures and a Wilcoxon test can be performed to determine the level of statistical significance of the result.

After the tests have been completed the index can show whether or not a procedure causes a significant disturbance in the animals' normal exploratory activity when confronted with a novel environment. If there is a significant disturbance however, it must not be assumed that the animals are necessarily displaying signs of pain, stress, or discomfort. It only demonstrates that the animals' normal activity pattern has been disturbed. The initial results using a tranquillizer and a stimulant change the behaviour and yield significant indices, but this does not imply suffering. Likewise the injection of thioglycollate broth, used when stimulating peritoneal macrophages, results in an enormous decrease in activity, but does not necessarily imply that the animal is suffering pain, because thioglycollate is known to have a sedative effect.

The value of the technique which has been described in this Report is that it provides an objective method for assessing the way in which various routine procedures affect an aspect of the normal repertoire of behaviour of a rodent. This makes it possible to compare equivalent procedures and select the one which is least disturbing. The technique is simple and the results are consistent with the known effects of two behaviour-modifying drugs. The fact that a DI can be tested statistically is also of great advantage.

When a procedure yields a non-significant DI it cannot be assumed that the animal does not suffer any pain or distress but it does suggest that the effect is not sufficiently persistent to disturb the subsequent behaviour of the animal. The DI method has the advantage that it is non-invasive, does not require elaborate equipment and provides an objective assessment of any lasting effects of different manipulations on rats and mice.

The observations on rats and mice reported here not only illustrate the utility of indices but provide guides to the choice of injection sites, volumes and vehicles, and peritoneal macrophage inducers.

Many of the observations it is realized, raise more questions than they answer. For instance, in the observations on dosing rats per os, what would be the effect of giving water at body rather than room temperature, or a saline or a sugar solution instead of water, or several small doses rather than one large one? Would the findings have differed if another strain of rat had been used? The facility with which such questions can be posed with expectation of an answer illustrates the utility of the method. It is able to generate objective measurements of disturbance and thus provide answers to questions where none were available before.

Finally, although the aim of the study was to validate the method of generating and statistical testing of the DI, the results presented may be of value to scientists using , or contemplating the use of, any of the routine procedures described.

6. RECOMMENDATIONS

It is recommended that whenever a procedure is to be used which is liable to produce pain, discomfort or suffering, a DI should be generated to determine the extent to which the procedure influences the animal's subsequent behaviour.

If the disturbance is found to be extensive, this must be taken into account and its possible interference with the validity of experimental results should be considered.

Finally wherever there is a choice between procedures, the least disturbing procedure should be identified and utilized.